Coding on the Playground

By Kristin Fontichiaro and Colleen van Lent

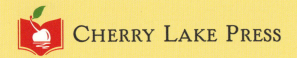

Published in the United States of America by Cherry Lake Publishing
Ann Arbor, Michigan
www.cherrylakepublishing.com

Series Adviser: Kristin Fontichiaro
Reading Adviser: Marla Conn, MS, Ed., Literacy specialist, Read-Ability, Inc.

Image Credits: ©AnnaliseArt/Pixabay, 4; ©OpenClipart-Vectors/Pixabay, 4, 12, 14, 16, 20; Various images throughout courtesy of Scratch;
Sound effects obtained from https://www.zapsplat.com

Library of Congress Cataloging-in-Publication Data

Names: Fontichiaro, Kristin, author. | Van Lent, Colleen, author.
Title: Coding on the playground / by Kristin Fontichiaro and Colleen van Lent.
Description: Ann Arbor, Michigan : Cherry Lake Publishing, 2020. | Series: Operation code | Includes index. | Audience: Grades 2-3.
Identifiers: LCCN 2019035731 (print) | LCCN 2019035732 (ebook) | ISBN 9781534159235 (hardcover) | ISBN 9781534161535 (paperback) |
    ISBN 9781534160385 (pdf) | ISBN 9781534162686 (ebook)
Subjects: LCSH: Scratch (Computer program language)—Juvenile literature. | Computer programming—Juvenile literature. | Playgrounds—Equipment
    and supplies—Juvenile literature.
Classification: LCC QA76.73.S345 F66 2020 (print) | LCC QA76.73.S345 (ebook) | DDC 005.1/18—dc23
LC record available at https://lccn.loc.gov/2019035731
LC ebook record available at https://lccn.loc.gov/2019035732

Cherry Lake Publishing would like to acknowledge the work of the Partnership for 21st Century Learning, a Network of Battelle for Kids.
Please visit http://www.battelleforkids.org/networks/p21 for more information.

Printed in the United States of America
Corporate Graphics

NOTE TO READERS: Use this book to practice your Scratch 3 coding skills. If you have never used Scratch before, ask a parent, teacher, or librarian to help you set up an account at *https://scratch.mit.edu*. Read the tutorials on the website to learn how Scratch works. Then you will be ready for the activities in this book! You will practice using variables, if/then statements, copying code to other sprites, using effects to change a sprite's look, and more! Find all the starter and final programs at *https://scratch.mit.edu/users/CherryLakeCoding*.

Table of Contents

Playground Trampoline

I'm Becca! Yesterday, the Parks Department installed a new trampoline at the playground. But can you keep a secret? I am scared I will fall if I jump on it.

Please help me write my code so I can bounce as high as the other kids. I got things started at *https://scratch.mit.edu/ projects/319364005*.

Pro Tip!

Scratch lets you see other people's projects and copy them to make them your own. My code contains **sprites**, backdrops, and the sound we will use in this book.

when 🚩 clicked

go to x: 25 y: -60

glide 1 secs to x: 25 y: 125

Going Up

My starter code places me on the trampoline and sets my **coordinates**.

Drag my sprite to the highest point I should jump. Scratch will automatically put the new coordinates in the Motion blocks.

Add the `glide 1 secs to x: 25 y: 125` Motion block to the code. This type of block will help me move smoothly.

Pro Tip!

When you want a sprite to go straight up and down, change the y-coordinate only. If you want it to go left or right, change the x-coordinate. To help you remember, think, "The y goes high!"

when 🚩 clicked

go to x: 25 y: -60

glide 1 secs to x: 25 y: 125

glide 1 secs to x: 25 y: -60

Coming Down

Test the code at least twice. Oh, no! I'm stuck high up in the air. Help me get back down to the trampoline.

Let's add another *glide* Motion block. Type in my starting coordinates, like this:

Test the code at least twice. Do I go up and down?

Pro Tip!

Good Scratch coders always test their code more than once. Sometimes, problems show up the second time you test your code.

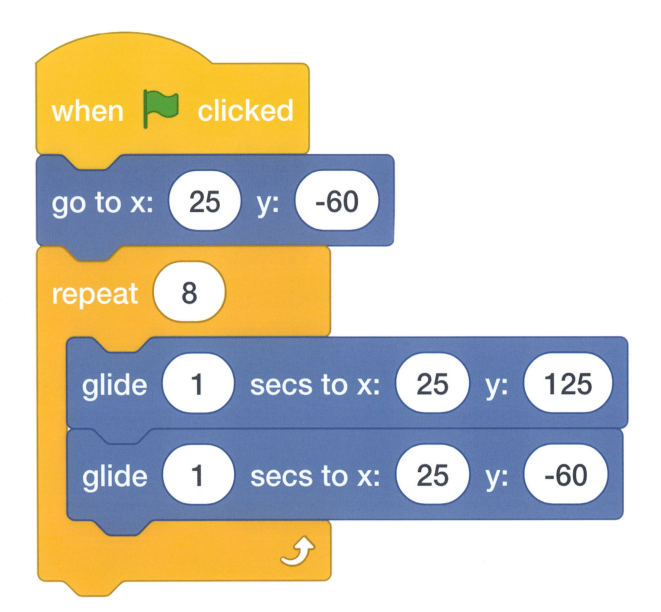

Repeating

I want to jump many times! I could add more *glide* blocks. But it would be easier to use the *repeat* Control block around the *glide* Motion blocks.

The number next to *repeat* says how many times Scratch should perform the blocks inside it. In this case, Scratch will repeat the *glide* Motion blocks eight times.

Test your code at least twice.

Pro Tip!

You want the *repeat* block to go around the two *glide* blocks. This can be tricky. Try dragging the *repeat* Control block right above the first *glide* Motion block. If it doesn't go around both blocks, try again.

Right-click to
show this menu ·······○   ○······· Costumes tab

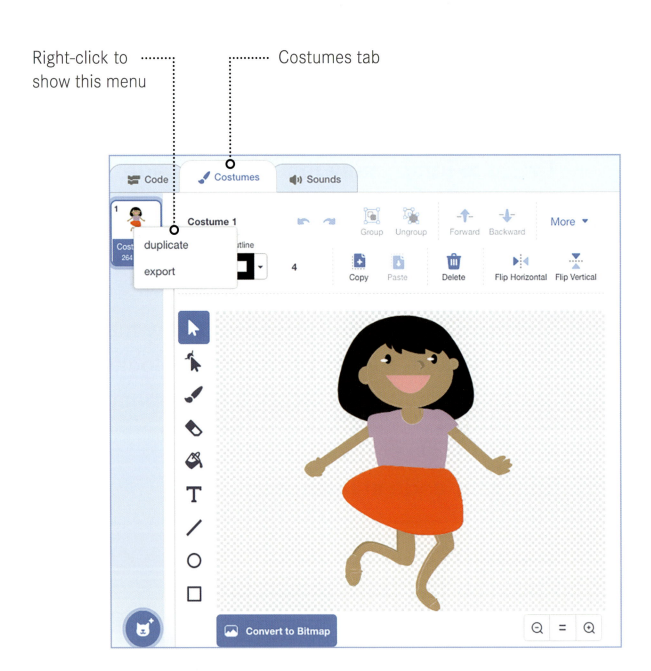

12

Duplicating a Costume

Let's change my body position when I jump. Some sprites have extra **costumes**, or different body positions. I only have one.

Let's try a **hack**. We will copy my costume and then flip its position so I look different. Click on the Costumes tab. Then right-click on the Costume 1 **thumbnail**. Click "duplicate" to make a copy.

Pro Tip!

If you see a blue box under my sprite that says "Convert to **Vector**," click it. That will make all the tools you need show up. The blue box will change to "Convert to Bitmap" and your screen will match mine.

Click "Flip Horizontal" to
look left instead of right ·······················

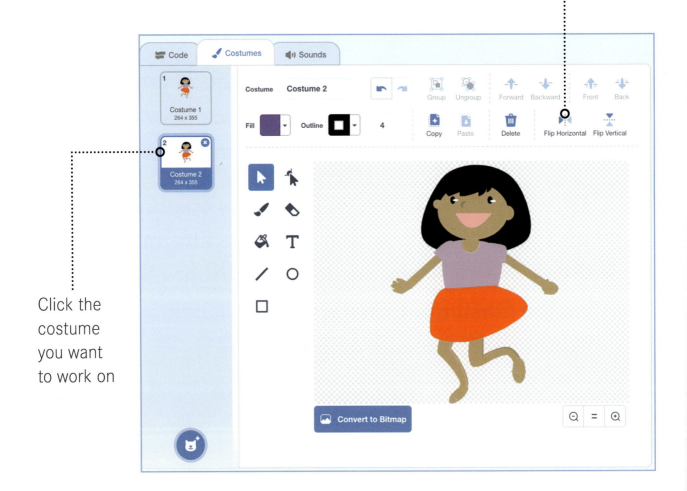

Click the
costume
you want
to work on

Flip Me!

Click on Costume 2. (That's the name Scratch gave our second costume automatically.)

Click "Flip Horizontal." Now I'm standing on my right leg, not my left.

Click between Costume 1 and 2 quickly. It looks like I am moving now!

If you want to see me turn headfirst, try "Flip Vertical."

Pro Tip!

Do you want to change anything else about my costume? Maybe add a flower to my shirt? Make sure you do that *before* you duplicate!

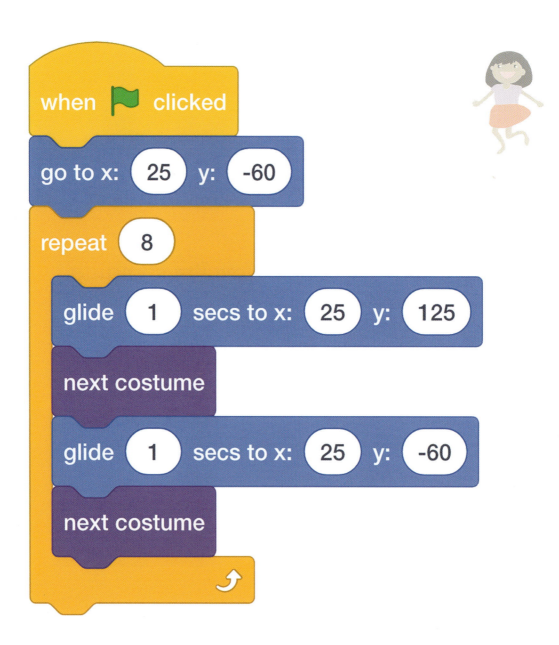

Adding the Next Costume Code

Now we have two costumes ready. Let's tell Scratch to change costumes.

Click back on the Code tab. We'll add the `next costume` Looks block to the code.

Try putting this block before or after your *glide* Motion blocks. Does it change what I look like?

Test your code at least two times.

Pro Tip!

The *next costume* **Looks** block rotates the costumes in the order they appear in the Costumes tab.

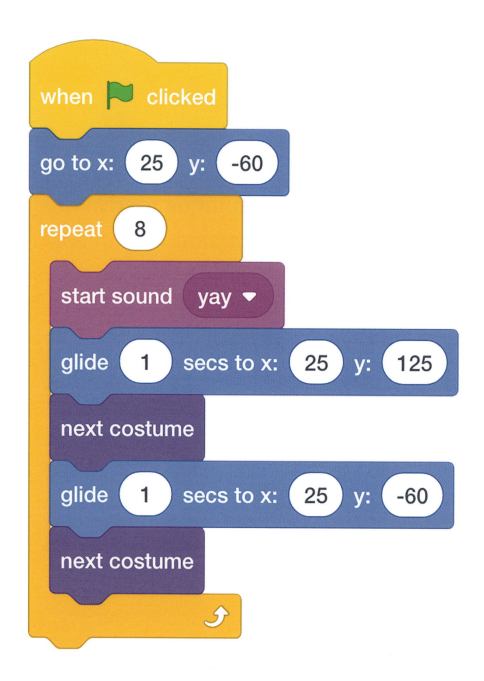

Adding Sound

Being on the trampoline is so much fun. I can't help but shout "Yay!" every time.

I **uploaded** a **custom** sound, called "yay," into Scratch. Can you help me put the [start sound ⬭] Sound block into my code?

Should I call out when I am up high or landing on the trampoline? Try putting the [start sound ⬭] block in different places.

Pro Tip!

Does your computer have a microphone? You can record your own voice and use that custom sound instead.

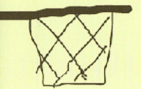

Eat healthy and move your body every day!

Keep Coding!

I had a blast jumping with you. Can you add more to our coding fun?

Here are some ideas:

- Change the speed or height of my jump.
- Duplicate the Trampoline sprite and add a new sprite to jump on it. Find a sprite with lots of costumes!
- Draw a different backdrop and use it instead of the playground.

What other ideas do you have?

Pro Tip!
To see our final code, please visit
https://scratch.mit.edu/projects/319257554.

Glossary

coordinates (koh-OR-duh-nits) numbers that tell the location of something

costumes (KAH-stoomz) a sprite's poses in Scratch

custom (KUHS-tuhm) unique or one-of-a-kind

hack (HAK) a workaround or another way of trying things with code when you can't do what you want to do

sprites (SPRYT) characters or objects in Scratch

thumbnail (THUHM-nayl) a tiny version of an image

uploaded (UHP-lohd-id) copied a file from your computer to Scratch, the Web, or another software

vector (VEK-tur) a kind of graphic that can be made bigger or smaller while still looking smooth on its edges

Find Out More

Books

LEAD Project. *Super Scratch Programming Adventure!*
San Francisco, California: No Starch Press, 2019.

Lovett, Amber. *Coding with Blockly.* Ann Arbor, Michigan:
Cherry Lake Publishing, 2017.

Websites

Scratch
http://scratch.mit.edu
Build your Scratch code online at this site.

Scratch Wiki
https://en.scratch-wiki.info
If you get stuck, ask an adult to help you look on this site for advice.

Index

About the Authors

Kristin Fontichiaro teaches at the University of Michigan School of Information. She likes working with kids on creative projects from coding to sewing to junk box inventions. She has written or edited almost 100 books for kids.

Colleen van Lent teaches coding and Web design at the University of Michigan School of Information. She has three cool kids and a dog named Bacon. She wishes she could touch her toes.